THE ROLE OF THE HEALTH AND SAFETY MANAGER

THE ROLE OF THE HEALTH AND SAFETY MANAGER

Andrew Watterson
De Montfort University, Leicester

and

Les Wright
British Geological Survey

The Health & Safety Series – Best Practice Management Reports
Series Editor: Melinda Taylor, South Buckinghamshire NHS Trust

Technical Communications (Publishing) Limited

ISBN 1 85953 015 X

Technical Communications (Publishing) Ltd.
PO Box 6
Hitchin
Hertfordshire SG5 2DB
England

Telephone/Fax: (01462) 437075

The publisher makes no representation, express or implied, with regard to the accuracy of the information contained in this publication and cannot accept any legal responsibility or liability for any errors or omissions that may be made.

The material contained in this publication constitutes general guidelines only and does not represent to be advice on any particular matter. No reader should act on the basis of material contained in this publication without first taking professional advice appropriate to their particular circumstances. The Author(s) and Publisher expressly disclaim any liability to any person who acts in reliance on the contents of this publication. Readers of this publication should be aware that only Acts of Parliament and Statutory Instruments have the force of law and that only courts can authoritatively interpret the law.

Printed in England

CONTENTS

FOREWORD

Brian M. Kazer, Past President of the Institution of Occupational Safety and Health

At last, a text which addresses a well identified gap in the literature:

- what does a health and safety (H&S) manager do?
- their benefits to an organisation.
- guidelines to their selection.

In addition to these practical and much needed pointers for prospective employers, the authors provide guidance both for employers, new and established health and safety managers (HSMs) on:

- what resources HSMs will need.
- what budgetary provision should be made.
- modern proactive H&S management

The need to address the proper place of a HSM in the organisational structure is particularly useful as lack of adequate consideration of this issue so often prevents full realisation of the potential contribution of a HSM.

The authors also give practical advice on the type and competence level of a HSM related to the size and degree of hazard of an organisation and illustrate this by a number of case studies.

In short, Andrew Watterson and Les Wright have filled a long standing gap in the market. More importantly they have produced essential practical guidance for prospective employers, newly appointed HSMs, and, dare one say it, for those with established HSMs to ask "are we getting the best value out of our present arrangements?".

The report is an essential read for all directors who have signed their company health and safety policy, for H&S recruitment specialists and for those considering a future career in health and safety.

PREFACE

Health and safety at work is an important subject in the modern workplace where many technical, legal and organisational demands are now being placed on large and small organisations to protect the well-being of their employees. As a result of the increase in importance given to legal, organisational and professional matters involving workplace health and safety, there has been a flood of publications, training courses, consultancies and advice offered to managers on the subject. Despite all this information, there is still confusion about the role of the health and safety (H&S) manager. This report answers the following questions in order to clarify the H&S manager's role:

- Who should carry out the tasks of managing health and safety in the workplace?
- What sort of tasks should they perform?
- To whom should they be accountable?
- How should they go about their work?
- What education and training should they seek?
- What qualifications and skills should an employer look for if they wished to appoint a H&S manager?
- What resources and facilities will the H&S manager need?

New H&S managers may often be faced with uncertainty and confusion about what they are meant to do, where they fit into an organisation and who they report to. Managing directors and human resource managers may also consider that they need a H&S manager but are uncertain about what duties such a manager should be given and what resources, responsibilities, training and information such an appointment would require.

This report looks at all these questions and provides some of the answers to help both employers select new H&S managers and those managers to start their work.

In the context of this report, the term 'H&S manager' is used to cover that individual who will have full-time or part-time responsibility for managing health and safety in the workplace. Other terms used to describe these people may variously include: health and safety officer, safety officer, health and safety adviser, safety adviser, safety practitioner and health and safety practitioner. All these terms cover the person responsible for managing health and safety in one form or another.

Some managers will simply 'advise' their seniors on good health and safety practice and on effective procedures, hazard removal and reduction strategies and organisation. But for health and safety to be managed effectively, the H&S manager's objectives must be fully integrated into the primary management objectives of the organisation. This is certainly how the UK's Health and Safety Executive (HSE) views the subject with its emphasis on successful health and safety management.

ABOUT THE AUTHORS

Dr Andrew Watterson is the Head of the Department of Health and Continuing Professional Studies at De Montfort University, Leicester, UK, where he is involved in a wide range of research and teaching on occupational health and safety subjects. He is also a Fellow of the Institution of Occupational Safety and Health and a Registered Safety Practitioner and has been involved in consultancy, education and research work on occupational health topics for twenty years.

Les Wright is the Health and Safety Adviser at the British Geological Survey, a component institute of the Natural Environmental Research Council. He is a part-time lecturer in occupational safety and health on certificate and diploma level courses at Nottingham Trent University and other commercial training centres in the East Midlands. He is a corporate member of the Institution of Occupational Safety and Health and a member of the British Health and Safety Society.

The authors can be contacted through the publishers.

CHAPTER 1

INTRODUCTION

This report aims to outline and examine the role of a H&S manager. The central role of such a manager is to develop a range of policies and practices best described as a health and safety culture. A positive health and safety culture is of major benefit to any organisation and can produce substantial cost savings.

The pressure on employers to manage health and safety effectively has increased over the last decade because of:

- the introduction of new legislation
- changing enforcement policies
- the threat of costly damages settlements.

The benefits from effective management can extend beyond compliance with legislation or reduction of personal injury risk, and give operational and economic benefits similar to those associated with quality assurance management programmes.

Your H&S manager should be able to use systematic health and safety management methods to eliminate the 'fire fighting' approach where problems are only tackled when they arise. This can be replaced with a strategy based on risk assessments which:

- identifies hazards
- anticipates adverse effects
- introduces control measures to eliminate or reduce risks through a realistic and achievable programme of management actions.

The health and safety culture comes from these developments. This culture then determines:

- the effective management of risks and hazards in the workplace
- the positive responses of senior management
- the ownership by employees of workplace health and safety issues.

The report is intended primarily for use by:

- new H&S managers looking for information about their new job
- managers who are aware of the need to enhance H&S management within their organisation
- managers who are considering the appointment of a H&S manager
- managers who wish to assess the effectiveness of existing health and safety management policies.

The skills necessary for a H&S manager to perform satisfactorily are likely to come from a combination of:

- education
- qualifications
- experience

both in H&S related disciplines and those reflecting the function of the organisation. This must include an awareness of the structure and management philosophy of the organisation.

The skills and experience so developed and applied can be termed **competence**. This is critical in the attainment of effective health and safety management. The importance of competence cannot be overstated and is indicated by the legally enforceable duty on employers from the Management of Health and Safety at Work Regulations 1992.

There is, however, no one approach which will serve the needs of all employment sectors. Case studies have therefore been included in the report to illustrate the various options available.

The health implications of health and safety are frequently under-emphasised, particularly where hazards are insidious, and the adverse health effects are not fully understood. These matters increasingly need addressing by the competent H&S manager as insurance claims have recently risen dramatically for occupational diseases with resultant productivity damage and increased costs to employers.

CHAPTER 2

WHY BOTHER WITH HEALTH AND SAFETY AT WORK?

The three main reasons for managing health and safety in the workplace relate to:

- **LEGAL factors**
- **ECONOMIC factors**
- **SOCIAL factors.**

Legal factors cover civil law and criminal law cases and incorporate much new European legislation. The report does not provide detailed guidance to the law. Appendix F does, however, list some of the key health and safety laws. There are two areas of employment law which influence workplace health and safety, these being **criminal** and **civil**. Criminal law consists of Acts and Regulations passed by Parliament, enforced by the Health and Safety Executive (HSE) (or in certain circumstances Local Authorities), under which non-compliance can lead to the issue of Improvement or Prohibition Notices (defined on page 6), or ultimately prosecution. Civil law allows individuals to seek financial settlement for damages where an injury has been experienced, usually through establishing negligence on the part of the employer.

Economic factors, especially commercial losses from accidents and ill-health, are a major element in reduced profitability in the 1990s. The economic effects of accidents and ill-health at work have been flagged up for many decades but the message has often failed to reach the senior management of commercial and other organisations. Now economists in the UK have provided details and case studies of how profitability and/or budgets and economic growth can be damaged very severely by poor health and safety management and the consequent staff and production losses (Davies and Teasdale 1994).

Social factors cover the reputation and credibility of organisations inevitably damaged and sometimes permanently damaged with ‘knock on’ user and consumer economic effects. For instance, in the marine and passenger transport sectors poor occupational health and safety records can damage if not destroy public credibility in a company or organisation, as can serious adverse environmental impacts. Ethical investors do not seek to locate funds in companies with poor occupational or related environmental health and safety records. Negative media coverage can seriously damage an enterprise and weaken the image and impact of any organisation.

CASE STUDY

At this stage, consider the following practical example of the problems created by an organisation neglecting legal, economic and social factors and failing to establish effective health and safety management roles, structures and procedures.

A small (i.e., less than 30 employees) manufacturing company required assistance with its health and safety management policy following a visit from the Health and Safety Executive which resulted in enforcement action.

The company manufactures steel-framed, faced particle-board office furniture using traditional engineering processes. The HSE had made an unannounced inspection and issued improvement notices for the following:

- Failure to make risk assessment relating to noise in production areas.
- Failure to produce records of examination and test of extract ventilation associated with de-greasing and welding plant.

The company could offer little evidence of an effective safety management policy, nor could it demonstrate an awareness of recently introduced regulations such as the Control of Substances Hazardous to Health (COSHH) and the Noise at Work regulations.

Consultants were engaged to meet the immediate needs of the improvement notice, and to develop the competence of existing management such that they could incorporate safety management functions into their primary duties.

The noise assessments required a specialised noise consultant equipped with sophisticated instrumentation to measure noise exposure in relation to action levels specified within the relevant regulations. The measurements of the extraction system, along with guidance on design improvements, and training to enable the company to set up an examination and test programme, were provided by a general consultant.

The work was completed during a five-day period, which included a seminar for the company's management, during which they were briefed on their health and safety duties, and provided practical training for the production foreman on measuring techniques. The company was able to demonstrate compliance with the notices when the HSE made a follow-up visit.

Economic, social and legal benefits

Had the company not been able to meet the conditions of the improvement notice, it is certain that the HSE would have considered further action, for example prosecution. This would have required further unscheduled time from various managers, and an adverse effect on production, including probable prohibition notices on the plant in which the regulations were being breached with subsequent adverse effects on production.

There were further benefits from the company's involvement with health and safety consultants in that they were made aware of the forthcoming regulations relating to the use of display screen equipment which presented significant new marketing opportunities for ergonomically designed computer furniture, which the company were able to exploit.

Social benefits accruing from the better working environment include:

- a happier workforce
- the avoidance of adverse publicity about the company damaging the hearing of its employees.

CHAPTER 3

WHY YOU NEED A HEALTH AND SAFETY MANAGER

There should be no conflict between effective health and safety management and organisational effectiveness if adequate resources and commitment are given to health and safety at work. Managers will save money, help avoid prosecutions and fulfil ethical requirements in their organisations if they deal with health and safety at work effectively. The following sections show how a H&S manager can minimise or avoid enforcement actions and reduce the costs associated with neglecting workplace health and safety.

AVOIDANCE OF ENFORCEMENT ACTION

Enforcement action results from intervention from the HSE or local authority enforcement officers where regulations have been breached. In practice this can be following an accident or routine inspection, and may relate to the presence of a particular uncontrolled hazard such as an unguarded machine, or more increasingly due to the absence of evidence that an employer has made risk assessments. H&S managers need to be concerned about enforcement action, which can be in the following forms. Associated costs have been indicated.

Improvement notices are issued to require employers to correct specific conditions or deficiencies in a facility or system of work, usually within a fixed time period. Costs include unscheduled staff time, plus costs of material, labour and any interruptions to normal operations to allow corrective measures.

Prohibition notices prohibit the use of the deficient facility, usually having immediate effect. The cost of rectifying faults will be similar to above but with more urgency; however, the loss of the facility will usually be most damaging, particularly where orders and other commitments to customers are likely to be affected.

Prosecution. If the breach of legislation is sufficiently serious the enforcement agencies may choose to proceed with criminal prosecution. The outcome of this could lead to manslaughter charges and custodial sentences for the person carrying the ultimate responsibility for health and safety in the organisation; to date, this has been limited to suspended sentences for the most serious offenders. Fines are more likely: the maximum at present being £20,000 per offence. Additional costs include court costs, legal representation, and staff time preparing for the defence. The cost of correcting the initial breach will still have to be met, and probably additional costs for similar activities not directly related to this action.

It is worth recognising that any visit from the enforcement agencies which does not result in either of the above is likely to result in some informal request for action, often requiring staff and resource allocation. Secondly, all the above actions are likely to result in follow-up visits, with the general principle being the further away from acceptable standards the enforcement agencies perceive your company, the more they will feel the need to intervene.

The costs associated with enforcement action are similar to accident costs in that there are direct costs such as fines and court fees, with additional less evident adverse effects including lost production time, unplanned resource commitment and staff time. There is little opportunity for insurance against these costs.

CHANGES IN ENFORCEMENT POLICY

The HSE have recently indicated that their workplace inspections will place less emphasis on the physical conditions of premises, shifting attention instead to interviewing senior managers in order to assess the effectiveness of health and safety management programmes. The basis for this is the recognition that the absence of effective management policies almost certainly leads to unacceptable standards. Hence forcing changes in the workplace will only be of limited value if the management attitude remains unchanged.

HOW CAN A HEALTH AND SAFETY MANAGER REDUCE THE IMPACT OF ENFORCEMENT ACTION?

The key to minimising the impact of visits from inspectors is for the company to achieve a high level of compliance with regulatory requirements. This will include both visible evidence of control measures such as personal protective equipment, safety signs and machine guarding. Equally important will be readily available evidence that health and safety has been managed in the workplace. For example, this could be achieved by:

- keeping records of risk assessments
- the production of management strategies for long-term action
- recording health and safety committee proceedings.

In general, the enforcement agencies would rather act in an advisory capacity, and will only take enforcement action where they feel advice has been ignored or conditions are so seriously hazardous that an accident is imminent. Inspectors will provide limited guidance over the telephone, and most likely send out free guidance literature or provide details of advisory publications that can be bought.

There will also be benefits from inviting the inspectors on to your sites because:

- this can remove the potential for unexpected visits
- this may allow for subsequent corrective actions to be scheduled within the company's primary management objectives.

Whilst such invitations might lead to enforcement action where conditions are unacceptable, it is more likely that advice will be provided. Once the inspector is confident that advice is being followed and an acceptable degree of compliance is evident, it is possible that the period between inspections will be increased.

Another way of developing a good relationship with the inspectors is for the H&S manager to become active in local health and safety professional associations as these will often include HSE and other enforcement officers as both members and speakers. This allows the membership to become aware of developments in the enforcement sector. In employment sectors where the hazards or control measures are specialised, the HSE may wish to work in partnership with employers to ensure the development of an effective health and safety management strategy.

AVOIDANCE OF THE COSTS ASSOCIATED WITH ACCIDENTS AND ILL-HEALTH

No company, organisation or individual normally sets out to have accidents but when they occur it is common to accept their inevitability. A fundamental principle of effective health and safety management is that accidents are predictable and therefore preventable. Much of the H&S manager's efforts will be directed at accident prevention and should be recognised as a cost saving exercise in the light of the costs of accidents, even allowing for expenditure on prevention strategies.

WHAT IS AN ACCIDENT?

An accident is an unplanned event resulting in injury or ill-health of people, or damage or loss of property, plant, materials or the environment or loss of a business opportunity. An accident is usually perceived as a single event resulting in instantaneous injury or effect.

This wider definition means managers need to recognise and assess the effects accidents will have on the business performance of the company or organisation. There may often be 'knock on' or indirect effects of accidents such as lost orders, damage to company profile or loss of customer confidence.

HOW MUCH WILL ACCIDENTS COST?

A major incident like the disaster on the Piper Alpha oil production facility is known to have cost 167 lives and approximately £2 billion; £746 million of which were direct insurance payouts. The more routine costs associated with most accidents in commerce and industry are less easily identified. It has recently been suggested that costs to the UK economy consist of:

- a total 27 million days lost per year in work-related injury and illness
- a sixty per cent increase in real-terms of employer liability insurance costs over the last decade
- a one hundred per cent increase in accident claims since 1985.

ESTABLISHING COSTS OF ACCIDENTS

In practice, the H&S manager will consider direct and indirect costs of accidents to organisations and hence establish real costs. If these are not currently monitored it will be necessary to set up a system to collect data on the following:

- Accidents resulting in personal injury
- Accidents resulting in damage to plant
- Near-miss events resulting in damage to plant
- Near-miss events not resulting in harm but indicating serious potential.

Summary of costs of an accident

Accident costs include direct and indirect cost, which can be depicted as an iceberg, with the direct costs representing the smaller visible section, but the larger submerged section representing the indirect or hidden costs. Recent studies have shown a relationship between each pound of direct costs to indirect costs as a ratio between 1:11 to 1:36, dependent upon industry sector.

Direct costs

These are often met by insurance cover and are concerned with a company's liability as an employer and occupier. Premiums are affected by the claims history, often leading to insurance companies placing conditions on the company to reduce risks. Direct costs may include legal costs, for example defence costs, claims and fines, and court costs: all of which will be difficult to predict.

Direct costs

Employers' liability claims and premiums

Legal costs (fees, fines and court costs)

Damage to buildings, plant, vehicles

Damaged product or material

Repairs

Sick pay and work-related disease costs.

Indirect costs

Interruption of business (lost time)

Investigation costs (staff time diverted from other tasks)

Additional staff cost (hired or replacement)

Loss of goodwill (workforce reaction)

Loss or damage to corporate image.

CASE STUDIES

1. A company building a supermarket over 12 months recorded all preventable accidents costing over £5 on the whole site in an 18 week period. 3,626 such accidents were recorded in this time at a direct financial loss of £87,507 and opportunity costs (mainly wages paid during no production) totalled an additional £157,568. For the 12 month building programme it was estimated that accident losses would total £700,000 or 8.5% of the tender price.

2. A 13 week study of a National Health Service hospital with 700 employees and an annual budget of £8 million recorded 1,232 accidents. Direct financial costs of the accidents totalled £48,000 plus and opportunity costs of approximately £51,000. In a full year the accident costs were estimated at around £397,000 or 5% of the hospital's annual running costs.
   (Source: HSE *The Costs of Accidents at Work.* 1993)

In both cases effective health and safety management would have made considerable savings for the organisation.

COSTS OF OCCUPATIONALLY CAUSED AND OCCUPATIONALLY RELATED ILL-HEALTH

Direct costs to employers of occupational ill-health have increased enormously in recent years with claims against employers increasing nearly three-fold between 1983 and 1988. Employers' liability claims are now estimated at £300 million a year.

CASE STUDY

A case of occupationally-related dermatitis requiring even minor medical treatment outside the workplace will incur significant costs from the employer through:

- lost time for medical treatments
- removal of the employee from exposure to agents thought likely to cause the disease and to allow the disease to clear up
- training and/or supervising the replacement
- personnel having to spend time and resources processing the above matters through the company
- costs of legal and insurance time in terms of responding to civil claims for damages due to the disease
- processing the paperwork for recording the case as a prescribed industrial disease (a prescribed disease is one listed in the schedule of the Social Security (Industrial Injuries) (Prescribed Processes) Regulations 1985).

A well-trained, well-educated and well-informed H&S manager should be able to identify potential health hazards at an early stage and devise systems to prevent exposures, hence saving these costs.

CAN HEALTH AND SAFETY MANAGERS HELP TO IMPROVE ECONOMIC PERFORMANCE?

As the importance of effective health and safety management becomes more apparent within various employment sectors – particularly market-driven ones – so it is becoming more common to find companies making specific efforts to assess the H&S management culture and performance of organisations with whom they will be working. This is so with suppliers and contractors too. This development is similar to that which established quality assurance standards as a prerequisite for successful trading, and is primarily aimed at reducing the liabilities or costs associated with poorly managed employers. The Management of Health and Safety at Work Regulations 1992 require employers to cooperate and coordinate health and safety measures where non-employees are likely to be affected by work or products of another party.

In practice the request for details may take the form of a questionnaire, examples of which have included the need to state the health and safety training undertaken by staff, along with details of relevant risk assessments and emergency procedures.

In some sectors it is becoming apparent that failure to demonstrate an acceptable degree of health and safety management will preclude, or at least seriously limit, the organisation's ability to operate in that sector.

EFFICIENCY GAINS AND PRODUCTIVITY BENEFITS THROUGH HEALTH AND SAFETY MANAGEMENT

Despite suggestions to the contrary, investment of resources primarily for risk reduction or compliance purposes can lead to significant efficiency gains. In the simplest case a workshop which is maintained in a clean and tidy condition not only meets the requirements of various legislation and reduces the risk of accidents, but the work carried out in it will be more efficient.

Where a risk assessment of an activity has identified hazards, it is often beneficial to subject the activity to detailed job safety analysis, a process where each stage is examined to devise measures to engineer out the hazardous elements. This detailed attention can result in alterations which not only improve safety, but lead to production benefits. Productivity benefits from health and safety management are now considered.

We have seen that the real costs associated with an unplanned event such as an accident can often be far greater than those at first visible. They include:

- injury costs
- sickness absence
- absenteeism
- lost time
- reduced productivity.

Lost time and reduced productivity are not only caused by absence of workers, but also by reduced effort and inefficient working practices caused by poor conditions and inadequately designed work methods and facilities. The hidden costs caused by this under-efficiency include:

- additional overtime
- over-staffing
- high staff turnover
- additional management and supervision time
- lower productivity

- product waste and damage
- plant damage and subsequent downtime.

Studies have been carried out to assess the financial benefit of effective health and safety management (Oxenburgh 1991), including the development of a productivity model which allows the cost of accidents, remedial measures and subsequent pay-back of these measures to be calculated. This allows the H&S manager to demonstrate the benefits of their role to other managers within the organisation in common (i.e., financial) terms.

The detailed and systematic evaluation of a task is necessary to identify deficiencies within it, thus allowing corrective measures to be introduced. Where a task has evolved over time, often including changes to plant or materials, there may not have been any obvious need to analyse the process, possibly allowing defects or inefficiencies to remain unchallenged.

The following case studies give brief details of the benefits which have been gained following modification primarily aimed at eliminating health or safety risks.

Substitution of a less hazardous substance leading to cost savings

A Swedish car manufacturer used ethanol and propanol-based solvents to clean metal and glass components within the interior of vehicles prior to delivery. This led to the exposure of operatives to high concentrations of fumes, often in excess of legally enforceable occupational exposure standards. The solvents were replaced by a cleaning solution formulated using soap, water and mildly abrasive chalk dust. This mixture had been used previously, but had been discarded decades ago following the introduction of solvents. The replacement cleaning mixture proved equally effective, and at a cost of 25% of the solvent-based product. It is worth noting that where alternative cleaning materials are introduced, particularly on large scale processes, there are usually significant environmental benefits, and probably reduced waste management costs.

Engineering out lifting, bending and poor posture leads to flexible staffing

There are numerous examples of industrial production activities which involve some form of manual handling likely to cause musculo-skeletal injuries due to poor workstation design. Having to manipulate heavy tools or workpieces will subject the arms, shoulders and lower back to loading likely to cause fatigue, and adversely affect productivity. These types of problem are solved by the application of ergonomic solutions, that is engineering solutions which lessen the impact of the task on the human body. In practice, this often involves simple, cost-effective measures such as the provision of a raised work support, or rearranging existing workstation equipment. More complex solutions might re-design work equipment to eliminate or redistribute its weight.

The primary benefit from reducing the manual force required to undertake a task will be a reduction in risk and subsequent injuries. A secondary benefit might be that a process previously only suitable for male operatives can now be safely carried out by females, presenting more flexible staffing opportunities for employers. Such benefits were gained by the United States Airforce following modifications it introduced to the size and shape of hand tools such as pliers, to allow for the different hand-span dimensions typical in females.

Getting it wrong – the introduction of technology to increase efficiency having the opposite effect

The introduction of new technology or processes aimed at achieving productivity gains through automation will not always lead to the desired reduction in health problems, as illustrated by the next case study.

A company wished to replace a manual paper-based accounting system with a computer-based system with the aim of increasing productivity. Due to the critical mistakes listed below, the change of process had the opposite effect.

- No consultation had taken place between the management and workforce, allowing suspicions to develop regarding the long-term objectives of the change – perceived as being staff reductions.
- Inadequate training had been provided on the use of the systems, resulting in an increase of stress amongst operators, likely to have an adverse effect on their mental and physical health.
- The computer terminals were deployed on tables used previously for traditional office functions, but unsuitable for continuous keyboard work due to differences in key dimensions necessary to allow the adoption of a comfortable posture.

Some of the operators had pre-existing neck and shoulder injuries, which had required medical attention, but had never led to sickness absence from work. These conditions were seriously aggravated by the poor workstations, leading to absence and subsequent productivity losses.

One of the most experienced staff suffered recurring problems, had several periods of extended sick leave, and eventually left the company. This led to additional costs associated with replacement (advertising, interviewing etc), and a general loss of efficiency in the office. It should also be noted that these circumstances could easily lead to a civil claim.

These difficulties could have been avoided by a programme which introduced the new technology following consultation, training, and selecting a new workstation. The cost of such actions have been estimated at equivalent of less than 250 employee salary hours.

CHAPTER 4

WHAT ARE HEALTH AND SAFETY MANAGERS?

If you are a managing director or human resources manager looking to appoint a H&S manager, it is hoped that you recognise and accept the benefits of such a post which have been outlined in the previous sections of this report. But what sort of a job description do you produce and what sort of a person do you look for? What role and function would they then have when they take up the post? The following sections aim to give you information and guidance on these aspects of health and safety management.

FULL-TIME OR PART-TIME?

One of the first considerations will be to establish how much time the H&S manager will require to be effective: this determines whether the post will be on a part-time or full-time basis. Ideally a H&S manager should be either full-time or with health and safety as the principal function of their work in most medium and large organisations. However, it will often be impossible for smaller workplaces to dedicate a full-time member of staff to deal with health and safety. In such circumstances, where there is a combination of a small workplace, few employees, very low risks and employees all on one site, then H&S managers may be part-time; they would still need to be trained and competent to carry out the tasks and management functions assigned to them.

WHAT IS A GOOD HEALTH AND SAFETY MANAGER?

A good H&S manager is someone with appropriate skills, knowledge, experience, training and education. The skills would include effective technical ability, organising ability and the capacity to communicate with those above and below the manager in the organisation. Depending on the nature of the organisation, the manager might also need to be a skilled trainer and be competent to assess and develop fire, emergency and disaster procedures.

Such a person would be able to use these attributes to identify deficiencies in the existing health and safety management structure or be able to establish an effective management structure in a new workplace. Through the establishment of effective procedures and practices and hence a positive health and safety culture, a good H&S manager would prevent accidents and ill-health in the organisation and reduce the number of safety and occupational disease problems.

Such a qualified person, given the appropriate support, will assist the organisation in ensuring the establishment of an efficient and effective workplace, and helping create a pleasant working environment.

COMPETENCE

Competence with regard to health and safety management is crucial to an effective health and safety management system. The H&S manager will require training to achieve and maintain the necessary levels of competence, and have the ability and expertise to provide training for managers, supervisors and workers as part of the in-house health and safety training programme.

Competence in health and safety terms is a combination of knowledge and experience. Knowledge of the elements listed below should enable the H&S manager to operate across a wide spectrum of disciplines found in the typical workplace:

- Legislation, acts and regulations, with emphasis on their realistic application in the relevant employment sector, the mechanisms of enforcement, and the role of civil courts in common law claims for damages.
- Health and safety management techniques including risk management and loss control, health and safety communication, accident prevention and safety monitoring techniques.
- A comprehensive understanding of workplace hazards, including their adverse effects on the human body, the methods of workplace monitoring and health surveillance and the application of relevant occupational exposure standards.
- The application of mechanical, electrical and construction engineering measures applied to control workplace hazards, including electrical safety devices, ventilation, personal protective equipment (PPE) and a range of hardware-based solutions described as safety technology.
- Aspects of human behaviourial sciences relevant to the motivation and communication of the health and safety policy within a mixed discipline workforce.
- Sources of information, advice and guidance, including British and international standards and toxicological data. Effective methods of communication.

Whilst there is no legal obligation for this knowledge to be gained by means of a formal qualification (there are examples of effective non-qualified health and safety practitioners), there are significant advantages from obtaining knowledge through the successful completion of a recognised National Vocational Qualification (NVQ) training course.

TRAINING

Occupational health and safety is a wide subject crossing the legal, scientific, behaviourial and technical disciplines. Without the guidance available from professional tutors this array of information can be daunting; however, when presented in a logical and complementary order such as that of a NVQ rated syllabus, the newcomer to the subject can more easily assimilate the necessary information.

There are a number of reputable training agencies, some of which are listed in the appendices of this report, which offer multiple levels of courses relevant to health and safety management. Employers are looking for staff with both the technical training, relevant educational background and good management skills and experience to take up such posts. There should be relevant technical training and education for holders of these posts. The National Examination Board for Occupational Safety and Health (NEBOSH) is recognised as the leading health and safety examining body in the UK.

- Good technical training entails either a NEBOSH Diploma or equivalent qualification by a reputable and rigorous training organisation or institution of higher/further education.
- Relevant educational background increasingly entails the possession of a first degree in occupational health and safety gained either through part-time or full-time study.

A number of part-time and full-time Masters of Science courses in occupational health and safety, occupational health and safety management, occupational hygiene and occupational and environmental health and safety are on offer. You will need to check that such courses are recognised as acceptable for membership by the leading professional bodies, for example the Institution of Occupational Safety and Health (IOSH). This is usually an indication of good quality. You will also need to decide whether you wish to specialise in certain branches of health and safety – for instance ergonomics or occupational hygiene – in which case choose a course dedicated to that subject or, if you wish to be a generalist – where possibly more job opportunities will exist with greater variety of work – choose an occupational health and safety/management course.

Managers with a degree in non-occupational health and safety subjects who wish to redirect their careers towards this subject can select either a post-graduate Diploma course or a Masters's conversion course. Do not forget that employers are increasingly looking for staff who can tackle occupational health and safety **and** environmental issues too. Such staff need relevant skills and experience.

These skills and experience should involve work in a range of activities if not a range of organisations with responsibility for identifying hazards, removing, reducing, monitoring and auditing those hazards and developing or assisting systems of health and safety management, procedures and policies.

TYPES OF QUALIFICATIONS AVAILABLE THROUGH OCCUPATIONAL HEALTH AND SAFETY TRAINING COURSES

Typical of effective and academically assessed courses would be:

- NEBOSH Certificate courses and other courses run by higher and further education institutions or approved commercial organisations for part-time H&S managers. Certificate level courses include introductory information on risk management, occupational hygiene, safety technology, health and safety law plus a practical examination to prove the ability to apply the knowledge in a workplace situation. Courses can be either intermittent day-release (1 day per week for 12-20 weeks) or block release (2-3 weeks).

- The NEBOSH Diploma or diplomas from further/higher education institutions approved by IOSH are generally accepted as the level of qualification necessary for a full-time H&S manager or part-time H&S manager in all but the smallest workplaces. Diploma courses develop further the basic subjects as presented at certificate level. The students are required to rely less on presented information, and are encouraged to research and apply information from libraries and other sources. Attendance at block release sessions is possible (4-6 weeks), but day-release over longer periods is more likely to produce better results due to the amount and complexity of the information required to attain this level of knowledge.

- First degrees are offered by a range of higher education institutions for part-time and full-time students who aim for a career in occupational health and safety. These allow the students to expand previous principles established at diploma level and integrate them more fully with other management disciplines.

- Taught Masters degrees which may be either 'conversion courses' for graduates in other subjects wishing to pursue careers in health and safety management or who wish to develop their skills in the field to an advanced level. Check that the course offers membership of IOSH on completion if you are not already an IOSH member.

- Specialist courses in occupational hygiene or ergonomics. Check with the relevant professional bodies for the latest courses and centres.

- Correspondence courses are an option for both certificate and diploma, but should only be considered where attendance of courses is impractical, as the personal contact with tutors and other students is a valuable part of the learning process.

Single-day and short duration seminars and training courses, often topic-specific, will be a useful way to gain knowledge on particular issues. A large number of commercial training centres and the recognised professional bodies regularly run programmes reflecting the recent developments within the field, for example the introduction of new health and safety regulations.

Training for health and safety managers will be an on-going process as many of the elements within the discipline are subject to revision and updating: for example the issue of new legislation, control methods and standards. This is recognised by IOSH which requires members to undertake continued professional development by attending training events in order to maintain full registered status.

CAREER PATHS FOR HEALTH AND SAFETY MANAGERS

Increasingly, organisations in the private and public sectors, in local government and hospitals, in consultancy and research, are requiring more and more professional H&S managers. This is linked to technological demands in the workplace and new legal standards generated within the UK and EC. H&S managers could therefore expect considerable job opportunities to be presented during their career. This could mean progression from an assistant H&S manager through to an Executive or Board position as a health and safety director in a large international corporation or major national body.

Case study on backgrounds of health and safety managers

Students occupying posts as H&S managers and attending a part-time degree course in H&S management familiar to the authors came from the following backgrounds:

- engineering
- engineering management
- human resource/management personnel
- fire service
- research and development
- retail
- legal.

SO WHICH OPTION MEETS YOUR ORGANISATION'S NEEDS?

Workplaces and organisations vary enormously in terms of the hazards which exist in them, the resources available and the management approaches, policies and philosophies. In one sense these variations are irrelevant to the health and safety role. There is no one job description for the work that H&S managers do. Templates will vary even within different sites of the same organisation, depending on the hazards, size of workplace and workforce and the particular state of the health and safety management, organisation and health and safety culture. See Table 4.1.

Table 4.1

Matching the H&S manager with qualifications by nominal workplace size and hazard criteria

Top comment indicates full or part-time H&S manager
Lower comment indicates recommended level of H&S qualification

Types of hazard	**Size of organisation (No. of employees)**		
	Small (Below 100)	**Medium (100-499)**	**Large (500 +)** See note 1
High - machinery, transport, chemical lab, microbiological hazards, radiation, lasers, repetitive actions, explosive and flammable gases, solvents etc.	Full or part time DIPLOMA	Full time DEGREE or DIPLOMA + MIOSH See note 2	Full time DEGREE or DIPLOMA + MIOSH
Medium - less hazardous substances, processes and systems than the above	Full or part time CERTIFICATE	Full or part time DIPLOMA + MIOSH	Full time DIPLOMA + MIOSH
Low - office work, human and social services	Part time CERTIFICATE	Part time CERTIFICATE	Full or part time CERTIFICATE

Notes:

1. Large = a site on several acres, of which several plants are spread across a geographically large area
2. MIOSH = Member of Institution of Occupational Safety and Health

WHEN SHOULD HEALTH AND SAFETY CONSULTANTS BE USED?

There are circumstances in which problems facing an organisation lie beyond the competence limits of the available H&S managers or advisers. In such cases the use of external consultants is a valid option, although care must be taken to ensure the right consultancy is engaged, and the services used to maximum benefit. When considering the use of consultants, it is vital to recognise that most duties under health and safety at work legislation rest unequivocally with the employer. Tasks can be delegated to others, for example consultants, but it remains the duty of the employer to ensure his legal obligations are met.

The following steps should be taken before deciding to use consultants:

- Identify a specific problem not effectively managed by existing arrangements.
- Analyse the problem to establish which elements within it can be tackled using in-house expertise.
- Consult a wide range of advisory material, including the publications and services available from the HSE and industry or trade associations.
- Consider whether the problem will occur routinely, suggesting the need for in-house solutions. This might be more cost-effective over the long term.

If following the above procedure indicates the need can not be met by existing in-house resources, it is likely that engaging a consultant will be beneficial. Independent expert opinion is valuable in supplementing in-house resources, and allows the development of alternative approaches to problem-solving. This can be particularly helpful where new processes, materials or technology are to be applied. Care must be taken if consultants are to be used to undertake statutory assessments to ensure that all parties are aware of such objectives at the outset of a contractual agreement.

The main areas of expertise offered by consultants are summarised below, although it should be noted that many consultancies will offer a range of services.

- **Technical or engineering.** Design and construction of technology-based control measures for a wide range of risk control systems, including exhaust ventilation, machinery guarding, noise and vibration control.
- **Management services.** Development and application of effective H&S management systems, including structure, organisation, policy development, performance monitoring, auditing, employee relations, training and communications.
- **Occupational hygiene.** Considers the effects of work and the working environment on health, including mechanisms of injury or ill-health, the assessment and control of airborne contaminants, and the selection and use of personal protective equipment. Occupational hygiene consultants will probably require the support of analytical laboratories.

- **Occupational medicine and nursing.** Services include diagnostics and nursing skills relevant to work-related injuries and ill-health, including health surveillance, medical examination or biological monitoring, first-aid support and training, stress management and general health education.
- **Ergonomists.** Assessment of tasks to determine solutions designed to meet anatomical and physiological needs of workers, particularly important in the fields of manual handling of loads and the design of workstations.

HOW DO YOU ASSESS A POTENTIAL CONSULTANCY?

Consultancies vary in size and capability from large multidisciplinary practices and university departments, through to small associations or individuals. The fees charged reflect this range, so it is important to select the most suitable consultancy for the problem at hand. General consultants charge between £200-£500 per day; specialist consultants charge up to £1,000 per day. The following questions should be asked of potential consultants:

- What previous work has been undertaken? Use business contacts, trade associations or local enforcement agencies to assess how good the service provided has been.
- What qualifications do the consultants have? NVQ level qualifications will be an indicator of ability, as will experience in specialised fields. Confirm membership of appropriate professional bodies such as recognised associations.
- What services are offered, particularly specialised areas such as noise, ergonomics etc.?
- How much will the work you need cost?

When a suitable consultancy has been selected, the next step is to ensure that the services they provide are exactly what you require by developing a task specification and outcome objectives. The specification must include details of the problem to be considered, including any preliminary measurements or other relevant data, your own interpretation of the legal considerations and an indication of what would represent a successful outcome.

The consultants should also be provided with relevant information about your organisation, including work patterns, plans, available resources and expertise. Budgetary guidelines and a time-frame may also be important.

HOW DO YOU ASSESS THE SUCCESS OR OTHERWISE OF A CONSULTANCY EXERCISE?

When the work has been completed, the following test should help assess the success or otherwise of the exercise:

- Have the aims of the specification been met?
- Have the results been provided in a form that you understand?
- Will the findings or recommendations make a positive contribution to the health and safety management of your organisation?

CHAPTER 5

THE ROLE OF HEALTH AND SAFETY MANAGERS

H&S managers may be full-time or part-time. They may be specialists in the field of health and safety, specialists in particular areas of health and safety or managers who have a role within the organisation to deal with health and safety. All H&S managers need proper and appropriate health and safety training, information and senior management support.

Depending on the size and nature of the organisation, the role and function of the health and safety practitioner will be defined. Many small workplaces or low risk workplaces will have a manager who takes on health and safety as a part-time role. Organisations will need to assess existing areas and levels of expertise among their staff to plan a coherent and coordinated approach to health and safety.

DEVELOPMENT OF THE MANAGEMENT STRATEGY

All H&S managers will need to ensure that their arrangements for health and safety at work are as outlined in the following key HSE document: *Successful Health and Safety Management*. The diagram below indicates the main elements in the recommended management strategy.

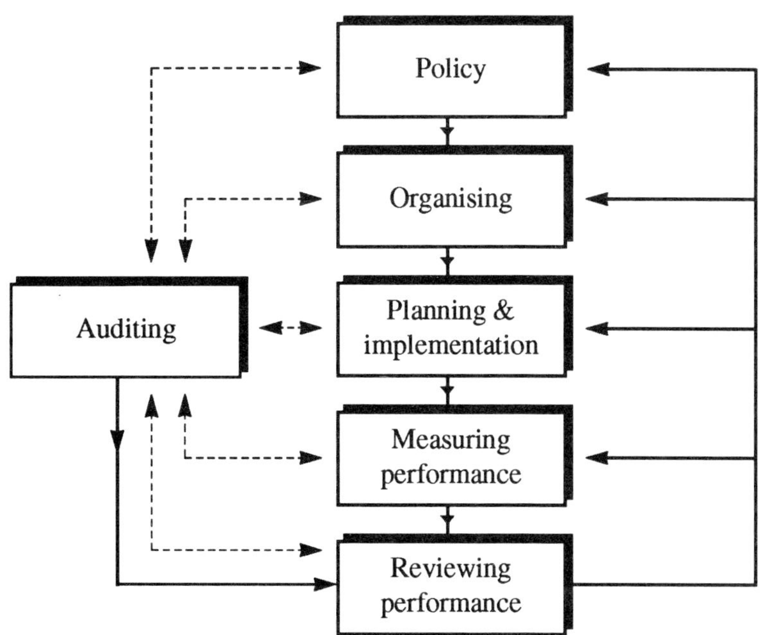

(Source: HSE *Successful Health and Safety Management* 1991)

ELEMENTS OF THE MANAGEMENT STRATEGY

The development of policy on health and safety which is recognised and supported throughout the organisation, and which seeks to minimise losses caused by unplanned events and optimise human resource development. Where these features are present the H&S manager will have a key role in ensuring the policy is updated to reflect external developments, such as in the introduction of new legislation.

Organising requires that certain key features must be present within the organisation to allow the effective management of health and safety to produce the maximum benefits. These critical elements include an appropriate management structure, unambiguous means of control, adequate resourcing and effective communication and coordination between all relevant sections of the organisation. Staff with key roles must be competent to undertake their functions and the structure must incorporate provisions for emergency conditions.

Planning requires the setting of health and safety objectives and the strategic progress towards the realisation of that plan. These objectives should aim to meet appropriate standards, but they must be realistic in that the objectives reflect the resources available and the likely influence of external or internal factors which might cause conflict with the planned objectives. In practice the H&S manager will utilise risk assessments as the device to identify the priority of tasks for inclusion in the plan. The H&S manager will often be faced with a number of hazards which require simultaneous assessment and control, whilst the available resources will only allow priority problems to be tackled immediately. It is essential that decisions regarding priorities are referenced to the overall plan.

Implementing requires that all measures established to achieve the plan objectives are actually present within the workplace, and that the plan is being adhered to.

Measuring performance of all aspects of health and safety will involve both active systems to confirm the compliance with specific requirements established in the plan and reactive systems which seek to analyse the impact of unplanned events such as accidents, ill-health or other losses, with the aim of minimising their effect. In practice, the H&S manager will develop and apply a number of performance monitoring initiatives, including the production of records, statistics and other information which can be fed back into the policy and planning elements of the health and safety management programme.

The effectiveness of all of the above elements of the management programme must be subject to regular **auditing and review.** This will seek to recognise failures before their impact is serious. Auditing is only effective if there is a corrective response to the detected failures. The H&S manager may be part of the audit team, although external auditing may also be effective. In practice the H&S manager might wish to make a formal report to the manage-

ment of the organisation on a regular basis, as would any other departmental manager. Organisations may choose to include information regarding health and safety performance in general company performance data.

KEY FUNCTIONS

The role of most H&S managers would include the following functions (from HSE *Successful Health and Safety Management,* 1991 and amended):

- to help formulate and develop policies
- to help structure and operate the health and safety organisation and systems
- to plan short-term, medium-term and long-term health and safety objectives and to prioritise those objectives
- to develop performance standards for the objectives set
- to implement the systems established on a day-to-day basis and to monitor and analyse policy and plans and performance
- to review health and safety performance in the organisation and audit the complete health and safety management system
- to maintain adequate information systems: both legal and technical
- to interpret and apply the relevant health and safety laws in the workplace setting
- to establish and, as appropriate, update organisational and risk control standards on hazards, equipment, systems of work and procedures
- to establish and maintain effective and appropriate monitoring and auditing systems
- to communicate advice in an effective and professional manner
- to identify hazards and risks and assess systems of work
- to investigate accidents, occupational diseases, near-misses and important health and safety problems
- to liaise with employees about health and safety
- to communicate with senior managers, production managers, maintenance staff and those purchasing goods and materials for the organisation
- to deal with insurance policies and claims
- to liaise with enforcement agencies ranging from fire officers, staff from environmental agencies, the National Rivers Authority, environmental health officers and health and safety enforcement officers from local authorities and HSE Inspectors.

DATA COLLECTION AND THE HEALTH AND SAFETY MANAGER

The health and safety practitioner should:

- collect
- store
- analyse

relevant data and, where appropriate, take action to amend employer policies and practices in the light of such data. This data is essential to the initial preparation of an employer's safety policy, the blueprint for health and safety action in the workplace, and in the promotion of the health and safety culture within the workplace.

THE IMPORTANCE OF COMMUNICATION

It is important that the H&S manager develops effective ways of communicating within the organisation. The communication will be two-way, and should include the following:

- data on hazards and risks
- the effect of these on specific areas of the workforce
- details on policy, procedures and systems of work
- reports on accidents and other performance indicators.

Oral presentation will be a key skill, particularly in training and in circumstances where a spoken instruction is likely to result in the desired response. Where such instructions are ineffective, written communication will formalise the instruction, and produce a record of key actions. Good quality report writing will often be necessary to maintain continued support from senior management.

Feedback from all areas of the organisation to the H&S manager will allow continuous assessment of the impact of his actions on the workplace.

CONSULTING THE WORKFORCE ON OCCUPATIONAL HEALTH AND SAFETY ISSUES

The law and good management practice dictates that the workforce is consulted on health and safety. This is achieved through meetings, multi-media education and training materials and the workplace health and safety committees. The H&S manager should also have regular meetings with other managers to look at health and safety problems. The culture necessary to achieve effective health and safety management requires good team management and team

building, good communications with managers above and below the H&S manager, and the development of clear health and safety procedures with the minimum of conflict.

Workforce involvement in the key stages of the development of a strategy to control risks will almost certainly increase the chances of that strategy being accepted and applied by the workforce as a whole. Specific areas in which workforce participation should be encouraged are:

- preparation to ensure compliance with legal requirements and the introduction of new requirements and technical advances.
- to identify training needs in the workplace.

HEALTH AND SAFETY MANAGERS – WHERE TO LOCATE THEM WITHIN THE ORGANISATION'S STRUCTURE

The H&S manager should have the position and status to advise the management and workforce with authority and independence. In practice this will depend on the type of organisation and particular management structure within which the H&S manager operates.

Good practice indicates that such managers should:

- have a direct reporting line to directors on health and safety policy
- have the authority to stop work activities where they are in contravention of agreed workplace practices and laws
- have responsibility for professional standards and systems
- be directly involved at the earliest stage of planning and purchasing in workplaces to pick up potential as well as existing health and safety problems
- be given all information relating to planning, purchasing, operations, revised and new and continuing systems of work
- where junior health and safety staff exist in the organisation, the senior health and safety manager would have line responsibility for such staff and be responsible for effective communication on health and safety with all staff above and below the senior H&S manager.

(Source: HSE *Successful Health and Safety Management* 1991)

RISK ASSESSMENT

Risk assessment is the starting point of any effective health and safety management strategy, and seeks to identify hazards and evaluate risks so that appropriate control measures can be introduced to manage the problem. Clearly, if the assessment is inaccurate, the control will be ineffective. Under-estimated risk can lead to adverse effects greater than those anticipated, whilst over-estimated risk leads to unnecessarily wasted resources.

The ability to make risk assessments is a practical test of competence and will require knowledge of both the health and safety implications and the process being considered – in many cases the H&S manager will be working closely with production managers, engineers or other specialists to ensure maximum benefit. The H&S manager would coordinate programmes of risk assessments and provide management with a schedule of action for risk reduction based on priorities. Risk assessment is a comprehensive topic in its own right and beyond the scope of this publication. However, a few key definitions may be helpful in understanding this important function:

(i) *Hazard*

Any condition or substance with the capacity to cause harm, injury, ill-health or loss of process. Examples include: toxic or corrosive substances, electricity, unguarded machinery, incorrect lifting, flammable or combustible materials.

(ii) *Risk*

The likelihood of the potential of a hazard being realised. Risk is a variable which will exist to some degree with any hazard, but can be reduced by the application of effective safety measures to an extent where the hazard can be said to be within control.

(iii) *Control measures*

Processes, procedures, facilities and equipment designed to maintain the safe condition in the presence of hazards. Examples include exhaust ventilation, machine guards, personal protective equipment, safety devices, and work methods, instruction, training and supervision.

The relationship between hazard and risk is the basis for successful maintenance of safe and healthy working. Where hazards are adequately controlled, risks are kept at a low level. If conditions change, for example in a fire or emergency or following the failure of protective devices, the risk increases.

As a practical example, consider a bottle of typist's correction fluid. Due to the presence of solvent within this product it can be said that this is a hazard. The labelling of the bottle will confirm this by including a warning pictogramme in accordance with packaging and labelling regulations. What is the risk associated with this product?

In normal use the correction fluid will probably be used for very short periods, probably too short to result in any significant exposure to the hazardous solvent, so we can say that the potential hazard will not be realised and the risk is low. Will it always be low? Could circumstances alter such that the risk is increased? The answer is yes. In the worst case scenario of someone correcting for long periods, perhaps in a poorly ventilated room, the exposure to the hazardous solvent may be significant and it can be said that the risk has increased. Other conditions which might increase the risk include accidental spills, fire or emergencies, or reaction with other substances.

The ability to make more subtle judgements of risk will become one of the most important skills the successful H&S manager will develop since it will be essential in allowing a number of competing problems to be ranked, and allow priorities to be established. For our purposes in developing a management strategy we will utilise a simple risk evaluation technique which assigns hazards and risks as low, medium or high. Giving each a numerical value and applying the formula below would allow a simple calculation to be made to enable the various risks in the workplace to be given a rating.

SIMPLE RISK ESTIMATION

Risk rating = hazard x risk.

Hazard values	High	3	Risk values	High	3
	Medium	2		Medium	2
	Low	1		Low	1

It may already be apparent that this simple risk index is limited and relies upon subjective evaluation of what is high, medium or low. More sophisticated models are available which take into account more conditions, but these are beyond the scope of this report.

PROVISION OF INFORMATION

One of the key roles of a health and safety practitioner is to provide information, both to initiate action within a risk reduction programme, and in response to requests for information regarding which regulations or standards apply to particular processes and how deficiencies can be remedied. A competent H&S manager will have a wide range of expertise to meet these enquiries, but more importantly will know where to find information which is not instantly available.

PROVISION OF TRAINING

All contemporary legislation includes requirements for employers to provide adequate health and safety training, which is seen as the key to competence, and subsequent risk reduction. The H&S manager would anticipate the training needs of the organisation in response to the strategy, and develop an appropriate training programme. This is likely to include in-house training, where the experience of the H&S manager and other suitably competent staff is utilised, and external training where the competence does not exist within the company.

MONITORING HEALTH AND SAFETY PERFORMANCE

For any management strategy to be effective it is necessary that:

- its progress is regularly monitored
- corrective action is taken where necessary.

The HSE have indicated the critical nature of this element in their guidance based on numerous examples of failed attempts to manage risk where the monitoring function has been absent.

H&S performance can be measured in a number of ways:

Negative performance indicators - accidents, ill-health, damaged plant and lost production time can be counted and calculated in the form of rates for comparison with either in-house norms or national/industry equivalents. Statistical analysis of these events can be useful where incident rates are sufficiently high to allow valid interpretation, but will be limited if data is scarce and infrequent.

Inspections, auditing, and sampling - there are a number of methodologies available which allow assessments to be made of the health and safety management of a process or premises following an inspection. Several points should be noted about inspections:

a. Regular inspections will allow comparative assessment, particularly when some form of rating or score is recorded.

b. Inspections are most effective when carried out in the company of a senior manager.

c. Inspections can be formal or informal depending on the normal operating regimes within the organisation, and should consider all aspects of the activity.

d. Safety auditing is a more comprehensive analysis of the health and safety performance, looking beyond the conditions and practices evident and into the management policy creating them. Auditing can be carried out by external agencies but should always involve the H&S manager.

An important part of the monitoring procedure requires the results to be presented to the management to allow corrective actions to be initiated where this is necessary. For this to be most effective, senior management must be fully conversant with the methodology employed, including differentiating between the desired and achieved performance levels.

INTERNAL ENFORCEMENT

Because HSE visits may be infrequent they can not be relied upon to monitor the level of compliance in the workplace, therefore it will be necessary for standards to be policed internally. The H&S manager can utilise the legislative framework as a template, allowing non-compliance to be detected in advance of an HSE inspection, enabling corrective action to be taken with minimum disruption to normal operations.

CHAPTER 6

WHAT RESOURCES DO HEALTH AND SAFETY MANAGERS REQUIRE?

To effectively manage health and safety it is necessary to have access to certain resources. These will of course vary depending on the type of organisation and its size, hazard profile and management structure. Resources should be sufficient to ensure effective information, communication and administration of health and safety matters within the workplace combined with sufficient clerical and secretarial support.

ADMINISTRATIVE/SECRETARIAL SUPPORT

This will be needed on either a full-time or part-time basis depending on the type of health and safety problems which exist and the size of the organisation. Systems to provide quality support, analysis and communication of health and safety data are vital to effective health and safety management. These systems are also central to the accountability of management structures and managers. They may also have to deal with matters of medical and legal confidentiality with the related security and data protection provisions operating.

INFORMATION RESOURCES

The key to the solution of any health and safety problem is access to information, particularly the relevant legislation and appropriate technical guidance publications. It is therefore essential that the health and safety manager has access to some form of reference library, including guidance notes, regulations, codes of practices and journals. If the organisation's technical library cannot be supplemented with a few key health and safety texts, it may be possible to have access to a local college or university library. Even allowing for access to a lending library, it will be necessary for the H&S manager to keep a few titles for quick reference. Keeping publications up to date can be a problem, and subscriptions to publishers which specialise in loose-leaf format guidance documents is recommended. A list of recommended titles is included in an appendix to this report.

Microfiche reader systems utilise large volumes of reference material in microfilm format, allowing easy access and limited copying facilities. Electronic media such as compact disc (CD) ROM systems allow access to vast quantities of data via computer hardware.

DESK TOP COMPUTER

An important tool in the production of quick but high quality communications can be achieved by means of word processing packages, also the storage and manipulation of data such as accident or assessment records using databases. This is particularly important in the absence of available administrative support. Where mobility around or between sites is important, a portable lap-top computer may allow flexible operations. Also, many environmental monitoring instruments are also greatly enhanced by the on-site presence of processing and display facilities of a personal computer.

FIELD EQUIPMENT

The role of the effective H&S manager will include a number of practical functions such as the measurement of various environmental hazards or the investigation of accidents. The following items of equipment are recommended for inclusion in a basic tool kit.

Camera

Pictorial evidence of conditions will be valuable during accident investigations. A compact 35mm camera should produce sufficient quality photographs without the need to master photographic techniques. Prints will be a useful addition to a report and slides can help in training and presentations. The logical development from stills photography is video recording which is particularly useful when undertaking manual handling assessments or other dynamic processes.

Audio tape recorder

Hand-held dictation recorders are useful for making notes during walk-round tours and inspections.

Environmental monitoring equipment

Although it should be noted that the H&S manager will not always be able to undertake environmental surveys to the level of an occupational hygienist or specialised consultant, the taking of instrumented measurements is a useful skill which can be a valid part of an environmental assessment.

Gas detector tubes

These devices are simple to operate and indicate the concentration of airborne contaminants such as solvent vapours and gases. The measurement is taken by drawing a precise amount of

air through a glass tube filled with crystalline chemicals which react by colour change and allow direct reading from the graduated scale printed on the tube. The sampling pump and a range of tubes for each contaminant needed to be monitored can be purchased for less than £200. Other occupational hygiene techniques such as dust sampling or detection by absorbent media will require analytical laboratory support.

Gas detection instruments

A number of direct reading electronic gas detectors may be appropriate if the presence of flammable or toxic gases is likely to be a hazard in the workplace. Single gas or multi-gas instruments can operate as an audible warning when critical concentrations are exceeded and store time series data via electronic loggers, allowing subsequent manipulation via computer software.

Temperature and humidity monitoring

A common requirement will be to investigate complaints regarding the environmental conditions within workplaces, particularly office environments where considerable concern is being shown about a number of adverse health conditions described as sick building syndrome. A simple direct reading temperature and humidity monitor with recording facilities, either electronic or chart drawing, will enable longer-term unattended measurements.

Light meter

As with the temperature/humidity indicator, the light meter will often be used in environments where complaints have been made about existing conditions. It will be useful when used in conjunction with recommended standards of lighting to eliminate hazards associated with incorrect lighting levels.

Noise meter

The accurate measurement of noise and the associated interpretation of such measurements fall within the specialised discipline of acoustics and are most effectively undertaken by specialist consultants. These services are, however, expensive and it will be cost-effective for the H&S manager to be able to make less sophisticated assessments of noise levels, if only to confirm that a problem exists, and that specialist help is required. Noise measuring instruments come in a range of accuracies, as reflected by a range in price from a few hundred pounds to in excess of ten thousand. A simple-to-operate sound level meter to a BS6698:1986/ ISO804:1988 Type 2 specification would be sufficient for preliminary surveys and should cost less than £500.

Air velocity indicator

A mechanical or solid-state thermal anemometer is a device to indicate air velocity – this is useful for assessing the performance of extraction ventilation systems used to control hazardous aerosols and room dilution ventilation systems. Multiplying velocities by the area of the ducting in which the measurements are taken will allow the calculation of volume flow rates, again a key performance indicator in ventilation systems. Instrumented readings can be supplemented by flow visualisation techniques using smoke tubes or pellets.

Small tools and accessories

A selection of small screw-drivers, pliers, torch and other small tools will provide a degree of independence from maintenance departments when using instrumentation or examining technical problems.

Personal protective equipment or PPE

Last but not least it is vital that the H&S manager is adequately equipped in terms of PPE for any environment or facility in which he is required to visit. The use of PPE is to be encouraged and the H&S manager must lead by example. For this reason the H&S manager may require a hard-hat, hearing protectors, safety footwear, high-visibility and weatherproof clothing, respiratory protective equipment and related necessary items.

COSTS

The costs of setting up and running an effective health and safety office include both capital and recurrent costs. Salaries, equipment and materials all need to be included in the budget. The figures below give a rough guide to such costs.

1.	Equipment. Environmental monitoring and consumables	£2,000
2.	Hardware and software systems to analyse technical and management data	£2,500
3.	Basic health and safety library	£500
4.	Staff costs	
	Junior health and safety manager	£10-15,000 / year
	Health and safety manager	£15-25,000 / year
	Senior health and safety manager	£25-30,000 + / year

Overheads would include membership of professional bodies and the costs of continuing professional development and training.

DEFENDING HEALTH AND SAFETY BUDGETS

The costs of employing and supporting competent health and safety staff will be covered by the savings that such staff will make in the production or operating budgets of organisations. Earlier sections have demonstrated the value of effective health and safety management in terms of economic, legal and social costs which may accrue to an organisation.

The costs of accidents, ill-health insurance, legal claims and replacement costs of equipment and employees should be included in managers' budgets. At present these costs are often covered and 'lost' centrally and managers may be penalised for expenditure to improve health and safety and prevent injury and disease. The inclusion of both types of costs in a manager's budget should make the benefits of good health and safety management manifest. This should help health and safety managers (hopefully!) defend their budgets.

CHAPTER 7

CASE STUDIES AND SKILLS NEEDED BY THE OCCUPATIONAL HEALTH AND SAFETY PRACTITIONER

EXAMPLES OF THE TYPE OF HEALTH AND SAFETY MANAGER, THEIR TRAINING, RESOURCES AND FACILITIES NEEDED

7.1 AN OFFICE WITH FEWER THAN 30 PEOPLE

Range of employees

Clerical, administrative staff, finance staff and a part-time human resources/personnel manager.

Hazards

Office environment – lighting, heating, ventilation, electricity, fire, VDUs, repetitive strain hazards, occupational stress, transport, chemicals in cleaning and photocopying, sick building syndrome.

Organisation

Managing director heads organisation with senior management support. Health and safety policy is needed plus risk assessment of various systems of work, materials and machinery.

Type of health and safety manager needed

Appointment of an existing administrator or human resources manager with part-time health and safety management responsibilities. The H&S manager would report to the managing director.

Consultants, if needed, could advise on details of office environments, Display Screen Equipment (DSE) regulations and sick building syndrome hazards.

Fire Brigade would offer assessments of fire precautions and hazards on visits.

Tasks

Development of policies and procedures.

Collecting data on accidents and near misses.

Ensuring company complies with health and safety legislation.

Health and safety communication.

Auditing, monitoring and reviewing.

Updating.

Training employees and managing director on health and safety requirements in the workplace.

Type of training needed

NEBOSH Certificate or equivalent training needed.

Resources

Office support and software support for the health and safety manager as part of the risk management functions.

Basic health and safety texts and laws.

7.2 A SMALL ELECTRONICS COMPANY EMPLOYING 100 PEOPLE WITH OFFICE AND WORKSHOPS

Range of employees

Technical staff, engineers, production workers, maintenance staff, office workers, managing director, personnel/finance staff.

Hazards

All the hazards of office work noted in 7.1.

Chemical hazards - solvents, solders and electronic materials.

Engineering hazards from various machines, oils, noise, dust and fumes.

Transport hazards, maintenance hazards, shift work.

Organisation

Managing director, finance manager and engineer head up the senior management team with sections dealing with production, sales, maintenance and new developments.

Type of H&S manager needed

Appointment of the company engineer in a part-time role but with proper training and high skills needed. The H&S manager would report to the managing director.

Tasks

Development of policies and procedures.

Assessing and monitoring risk assessments of chemical and engineering processes.

Liaising with HSE Inspectors and offices.

Attending industry-based meetings and courses on hazards of electronics industry, occuptational hygiene measurements of fumes, dust, noise and vibration.

Collecting data on accidents and near misses.

Ensuring company complies with health and safety legislation.

Health and safety communication.

Auditing, monitoring and reviewing.

Updating.

Training employees and managing director on health and safety requirements in the workplace.

Consultants could undertake specialist monitoring of hazards, carry out assessments of procedures and devise engineering solutions to hazards identified.

Type of training needed

NEBOSH Diploma or equivalent.

Resources

Office support.

Suitable software to run monitoring and auditing system of hazards.

Occupational hygiene monitoring equipment.

Journals and books on industry-based hazards and laws.

7.3 A SMALL MICROBIOLOGICAL LABORATORY EMPLOYING 15 PEOPLE

Range of employees

Manager, microbiologists, laboratory technicians, part-time cleaners and contract maintenance staff, secretary.

Hazards

Office hazards as 7.1.

Microbiological hazards including various pathogens and carcinogens.

Fume cupboards.

Type of H&S manager needed

Microbiologist working part-time could cover the health and safety management needs and should report to the manager.

Tasks

Development of policies and procedures.

Assessing and monitoring risk assessments in the laboratories including Control of Substances Hazardous to Health (COSHH) Regulations assessments.

Liaising with HSE Inspectors and offices.

Attending meetings and courses on laboratory hazards.

Occupational hygiene measurements of fumes and pathogens.

Monitoring the condition of laboratory equipment and its effectiveness.

Collecting and analysing data on accidents and near misses.

Ensuring company complies with health and safety legislation.

Health and safety communication.

Auditing, monitoring and reviewing.

Updating.

Training employees and management on health and safety requirements in the workplace.

Consultants could undertake specialist monitoring of hazards, carry out assessments of procedures and devise engineering solutions to hazards identified.

Type of training needed

NEBOSH Diploma.

Resources

Office support.

Suitable software to run monitoring and auditing system on hazards.

Occupational hygiene monitoring equipment.

Journals and books on laboratory hazards and laws.

7.4 LARGE WAREHOUSE AND DISTRIBUTION CENTRE WITH 100 EMPLOYEES

Range of employees

Managing director, managers, office staff, warehouse workers, maintenance workers, transport drivers - lorries and fork-lift truck drivers.

Hazards

Office hazards as 7.1.

Transport hazards - both internal and external.

Maintenance hazards.

Hazards of warehousing in terms of moving loads, stacking etc.

Organisation

Managing director.

Type of H&S manager needed

Part-time H&S manager.

Tasks

Development of policies and procedures.

Collecting data on accidents and near misses.

Ensuring company complies with health and safety legislation.

Health and safety communication.

Auditing, monitoring and reviewing.

Updating.

Training employees and managing director on health and safety requirements in the workplace with particular reference to warehousing and related hazards.

Type of training needed

At least a NEBOSH Certificate in Occupational Safety and Health or equivalent.

Resources

Office support.

Suitable software to run monitoring and auditing system on hazards.

Basic monitoring equipment of fumes from vehicles and lighting.

Journals and books on general health and safety and hazards of warehousing.

7.5 LARGE MANUFACTURING COMPANY WITH A HEAD OFFICE BASED AT ONE SITE AND THREE OTHER SITES - 500 EMPLOYEES

Range of employees

Technical staff, engineers, production workers, maintenance staff, office workers, managing director, personnel/finance staff.

Hazards

All the hazards of office work noted in 7.1.

Chemical hazards - solvents.

Solders and electronic materials.

Engineering hazards from various machines, oils, noise, dust, fumes.

Transport hazards, maintenance hazards, shift work.

Organisation

Managing director, finance manager and engineer head up the senior management team with sections dealing with production, sales, maintenance and new developments.

Type of H&S manager needed

Full-time H&S manager for the company and part-time H&S managers on each site, preferably the engineers or personnel managers taking on the part-time post.

Tasks

Development of policies and procedures.

Assessing and monitoring risk assessments of chemical and engineering processes.

Liaising with HSE Inspectors and offices.

Attending industry-based meetings and courses on hazards of electronics industry.

Occupational hygiene measurements of fumes, dust, noise and vibration.

Collecting data on accidents and near misses.

Ensuring company complies with health and safety legislation.

Health and safety communication.

Auditing, monitoring and reviewing.

Updating.

Training employees and managing director on health and safety requirements in the workplace.

Consultants could undertake specialist monitoring of hazards, carry out assessments of procedures and devise engineering solutions to hazards identified.

Type of training needed

NEBOSH Diploma or equivalent or degree for the full-time health and safety manager.

NEBOSH Certificate or equivalent for the part-time managers.

Resources

Office support.

Suitable software to run monitoring and auditing system on hazards.

Occupational hygiene monitoring equipment.

Journals and books on industry-based hazards and laws.

SKILLS NEEDED BY THE OCCUPATIONAL HEALTH AND SAFETY PRACTITIONER

Recent developments in the UK in the field of NVQs have led various professional bodies towards establishing guidelines for NVQs in health and safety. At the time of writing, nothing had been agreed formally. However, the work of the Occupational Health and Safety Lead Body in developing the document *Draft Standards of Competence for Occupational Health and Safety Practitioners* (see references, page 51) has produced some useful indicators of the type of standards which might apply.

The Occupational Health and Safety Lead Body has been set up, at the instigation of the Government and agencies concerned with national vocational qualifications, to examine the vocational qualifications and related standards of competence needed by those involved in the field. The professions involved in the lead body are principally health and safety practitioners, occupational health nurses, occupational hygienists and occupational physicians.

The following standards are derived from *Draft Standards of Competence for Occupational Health and Safety Practitioners* and are listed simply as a useful indicator of what employers and practitioners may be looking for in the competent practitioner. They do not have any legal, educational or professional standing at the time of writing and several categories have already been amended and further revised. The lists are not exhaustive and skills needed may vary from job to job.

The precise vocational qualification levels at which health and safety practitioners will work has not yet been established.

Level 2 are not health and safety practitioners but support health and safety practitioners. Level 3 tends to cover the suggested qualifications needed by a general 'trainee' health and safety practitioner but could also apply to someone with a health and safety practitioner role working alone in an organisation. Level 4 tends to cover a fully competent professional health and safety practitioner. Level 5 would cover a health and safety practitioner working in senior or high middle management with responsibility for a team of health and safety practitioners.

THE SKILLS NEEDED BY A LEVEL 2 OCCUPATIONAL HEALTH AND SAFETY PRACTITIONER MIGHT INCLUDE:

- the ability to promote and support the establishment and maintenance of health and safety systems
- the ability to assess the requirements of risk control
- contributing to the design and implementation of risk control

- promoting and supporting the continuous development of a health and safety culture
- promoting and supporting compliance with, and continuing raising of practice above, legal standards.

THE SKILLS NEEDED BY A LEVEL 3 OCCUPATIONAL HEALTH AND SAFETY PRACTITIONER MIGHT INCLUDE:

- helping to control risks in defined areas
- contributing to drafting and communication of an organisation's health and safety policy and other documentation
- promoting and supporting the establishment and maintenance of health and safety systems
- contributing to assessing risk control requirements
- designing and implementing certain risk controls
- promoting and supporting the continuous development of a health and safety culture
- promoting and supporting compliance with, and continuing raising of practice above, legal standards
- in those workplaces without a training officer or training department, contributing to the training of individuals and groups on health and safety at work
- creating, maintaining and enhancing positive relationships on health and safety issues
- providing information and advice for effective action on health and safety aims and objectives of the organisation
- contributing to, and keeping up with, advances in the field of occupational health and safety.

THE SKILLS NEEDED BY A LEVEL 4/5 OCCUPATIONAL HEALTH AND SAFETY PRACTITIONER MIGHT INCLUDE:

- helping to control risks in defined areas
- contributing to drafting and communication of an organisation's health and safety policy and other documentation
- promoting and supporting the establishment and maintenance of health and safety systems
- contributing to assessing risk control requirements
- designing and implementing certain risk controls
- promoting and supporting the continuous development of a health and safety culture

- promoting and supporting compliance with, and continuing raising of practice above, legal standards
- identifying organisational training and development needs of health and safety at work
- identifying current health and safety competence of individuals
- creating, maintaining and enhancing positive relationships on health and safety issues
- providing information and advice for effective action on health and safety aims and objectives of the organisation
- contributing to and keeping up with advances in the field of occupational health and safety
- exchange information to solve problems and make decisions
- evaluate achievement of outcomes against health and safety learning and general objectives
- contribute to the implementation of necessary changes in services, products and systems with regard to health and safety matters.

CHAPTER 8

PROGRAMME OF ACTION

We will assume you are the newly appointed H&S manager of a company or organisation which has recognised the need to improve its health and safety management, but has an under-developed health and safety culture. It is important to recognise that health and safety management is a very broad field and changes on a large scale are unlikely to be achieved overnight. Effective management of this range of problems is only likely to be achieved if a strategic approach is adopted.

The starting point is to ask the following key questions:

- What is the policy of the organisation towards health and safety?
- What are the health and safety management problems in the organisation?
- What resources are available to eliminate or control these problems?
- In what order should the problems be tackled?

The following chapter provides practical guidance on the development of a programme of action to establish an effective health and safety management strategy. The steps necessary will include both desk-top and on-site activity.

ESTABLISH ROLES, ORGANISATION AND ACCOUNTABILITY

Before the H&S manager can expect to operate within an organisation, or hope to influence its activities, it will be necessary to establish the exact role of the H&S manager, and the managerial structures within which he/she will operate. Equally important is the establishment of a clear chain of command: including clear indications from senior managers to the extent of the H&S managers' authority, and the limit to which they might operate without direct line management approval. In reality, some of the health and safety manager's functions will result in some degree of conflict with workers or other management, particularly where the advice or instruction given is seen as unwelcome and uninvited. Such situations can only be effectively managed with full support from senior line managers.

EVALUATE EXISTING POLICY

The existing policy of the company or organisation should be assessed to establish that it meets minimum legal requirements and whether it is likely to promote effective health and

safety management. Where there is no health and safety policy statement, any relevant managerial instructions or procedures should be compiled to allow the future development of a policy statement.

ESTIMATE CURRENT ACCIDENT AND ILL-HEALTH COSTS

It will often be easier to maintain support for health and safety initiatives, particularly those requiring the use of resources, by indicating potential savings against existing costs. In the absence of detailed accident costs it should be possible to take accident records, perhaps based on entries in the accident book or any other available record, and apply value in terms of direct and indirect costs (see chapter 3). Where sufficient data exists it may be possible to make direct comparison with expected accident industry sector statistics.

ESTABLISH EXISTING RESOURCES

Assess the health and safety management expertise currently available within the organisation. You might be surprised to realise how many functions include significant amounts of health and safety which can be utilised within a health and safety management programme. Examples include maintenance engineers, training officers, laboratory managers, personnel managers, occupational nurses and quality assurance specialists.

IDENTIFY HAZARDS AND EVALUATE RISKS

Once you have a reasonable idea of the facilities and expertise available, you can begin to establish what health and safety problems exist in the organisation. Whilst some of the hazard identification and risk evaluation can be carried out as a desk-top exercise, the only way to truly assess the condition and attitude to safety in the workplace will be to undertake inspection tours. These are often more productive when accompanied by a representative of the area being inspected, and often achieve a higher level of local support if prior warning is given.

These hazard identifying tours should not be confused with safety inspections or auditing exercises, but should seek to do the following:

- Identify existing hazards
- Evaluate risks
- Identify existing control measures
- Identify any apparent deficiencies in control measures.

ESTABLISH STRATEGY

The hazard identification tours should lead to the production of a list of problems. These should be ranked according to the potential adverse effect, and fed into the health and safety management strategy. This would enable the H&S manager to:

- Identify in-house resources which can assist in the control of hazards.
- Rank the problems to ensure priority treatment for the most hazardous.

Each risk management task should be classified and assigned a date by which key objectives have to be met. In addition, the individuals responsible for a particular phase of the exercise must be identified.

MONITORING ITS EFFECTIVENESS

As with any area of management, it will be important to periodically monitor the progress of the health and safety management systems. Most activities will involve a number of interested parties and much can be achieved by informal working groups. When necessary, a health and safety committee can formalise issues and offer an obvious means of ensuring that senior management are aware of the progress or otherwise of health and safety issues.

REFERENCES

Davies N.V. and Teasdale P. (1989) *The Costs to the British Economy of Work Accidents and Work-Related Ill Health*, HSE Books, London.

Health and Safety Executive (1991) *Successful Health and Safety Management*, HMSO, London.

Health and Safety Executive (1993) *The Costs of Accidents at Work*, HMSO, London.

Health and Safety Executive (1992) *Selecting a Health and Safety Consultancy*, HMSO IND(G) 133 (L), London, 12/92 C500.

IOSH (1993) *Draft Standards of Competence for Occupational Health and Safety Practitioners*: Consultative Document, IOSH, Leicester.

Oxenburgh M. (1991) *Increasing Productivity and Profit Through Health and Safety*, CCH International.

APPENDICES

A. SOURCES OF INFORMATION

HSE Information Centre
Broad Lane
Sheffield
S3 7HQ
Tel: 0742-892345
Fax: 0742-892333
Free leafletline: 0742-892346

The Royal Society for the Prevention of Accidents
Cannon House
The Priory Queensway
Birmingham B4 6BS
Tel: 021-200-2461

B. PROFESSIONAL AND CONSULTANCY BODIES

INSTITUTION OF OCCUPATIONAL SAFETY AND HEALTH (IOSH)
The Grange
Highfield Drive
Wigston
Leicester LE18 1NN
Tel: 01162-571399

NATIONAL EXAMINATION BOARD FOR OCCUPATIONAL SAFETY AND HEALTH (NEBOSH)
NEBOSH House
Newton Lane
Wigston
Leicester LE18 3PP
Tel: 01162-888858

INSTITUTE OF OCCUPATIONAL HYGIENISTS
Georgian House
Great Northern Rd.
Derby DE1 1LT
Tel: 0332-298087

BRITISH OCCUPATIONAL HYGIENE SOCIETY
Georgian House
Great Northern Rd.
Derby DE1 1LT
Tel: 0332-298101

THE ERGONOMICS SOCIETY
Devonshire House
Devonshire Square
Loughborough LE11 3DW
Tel: 0509-234904

INDEPENDENT SAFETY CONSULTANTS ASSOCIATION
c/o Hinton and Higgs
The Firs
Macham Rd.
Abingdon
Oxon OX14 1AA
Tel: 0235-524228

INSTITUTE OF ENVIRONMENTAL HEALTH OFFICERS
Chadwick House
Rushworth St.
London SE1 0QT
Tel: 071-928-6006

INSTITUTE OF ACOUSTICS
PO Box 320
St Albans
Herts AL1 19Z
Tel: 0727-48195

INSTITUTE OF ELECTRICAL ENGINEERS
Savoy Place
London WC2R 0BL
Tel: 071-240-1871

INSTITUTE OF MECHANICAL ENGINEERS
1 Birdcage Walk
London SW1 9JJ
Tel: 071-222-7899

INSTITUTE OF OCCUPATIONAL MEDICINE
8 Roxburgh Place
Edinburgh EH8 9SU
Tel: 031-667-5131

ROYAL COLLEGE OF NURSING
Society of Occupational Health Nursing
RCN North Western Area
18 Fox St
Preston
Lancs PR1 2AB
Tel: 0772-252891

SOCIETY OF OCCUPATIONAL MEDICINE
6 St Andrew's Place
Regents Park
London NW1 4LB
Tel: 071-486-2641

C. EDUCATIONAL/TRAINING INSTITUTIONS OPERATING HEALTH AND SAFETY COURSES

The best sources of information on these elements are available through either the Institution of Occupational Safety and Health or the National Examining Board on Occupational Safety and Health. The addresses and telephone numbers of both these bodies are listed in Appendix B.

IOSH will be able to tell you which higher education institutions offering qualifications in occupational health and safety are recognised for the purposes of IOSH membership. The colleges doing this and the courses on offer are changing rapidly at present.

Universities offering first degree, post-graduate Diploma or Masters courses in occupational health and safety currently include:

Aston University

Glasgow Caledonian University

Leeds Metropolitan University

Loughborough University

Nottingham Trent University

University of Glamorgan

University of Greenwich

University of Paisley

University of South Bank

NEBOSH can provide you with a list of colleges in the further/higher education sector which offer the NEBOSH Diploma and Certificate courses – such colleges are spread out across the country.

Some institutions, either in further/higher education or commercial training bodies, also offer courses at Certificate and Diploma level which are recognised by IOSH.

D. KEY TEXTS AND RECOMMENDED READING

[* indicates essential addition to H&S manager's library]

Successful Health and Safety Management.
Health and Safety Executive (HSE)
HSG (65) HMSO 1991 ISBN 0 11 885988 9

**Essentials of Health and Safety at Work.*
HSE HMSO ISBN 0 11 885445 3

Increasing Productivity and Profit Through Health and Safety.
Maurice Oxenburgh
CCH International 1991 ISBN 1 86264 264 8

**Management of Health and Safety at Work.*
Approved Code of Practice - Management of Health and Safety at Work Regulations 1992.
HSE, HMSO. 1992 ISBN 0 11 886330 4

Selecting a Health and Safety Consultancy.
HSE IND(G) 1992 133L (Free leaflet)

Five Steps to Successful Health and Safety Management.
HSE IND(G) 1992 132L (Free leaflet)

Tolleys Health and Safety at Work Handbook.
Malcolm Dewis
Royal Society for the Prevention of Accidents/Tolley Publications. 1993 ISBN 0 85459 639-9

The Handbook of Health and Safety Practice.
Jeremy Stranks
RoSPA/Pitman 1993 ISBN 0 273 03202 X

Health and Safety - The New Legal Framework.
Ian Smith, Christopher Goddard, Nicholas Randall
Butterworths 1993 ISBN 0 406 02299 2.

Monitoring for Health Hazards at Work. Second edition.
Indria Ashton and Frank S Gill
Blackwell Scientific Publications 1992 ISBN 0 632 02984 6.

E. ACRONYMS

BEBOH - British Examination Board for Occupational Hygienists
BHSS - British Health and Safety Society
BOHS - British Occupational Hygiene Society
COSHH - Control of Substances Hazardous to Health
EC - European Community

EU - European Union
H&S - Health and Safety
HSC - Health and Safety Commission
HSE - Health and Safety Executive
IAC - Industrial Advisory Committee of HSC
IOSH - Institution of Occupational Safety and Health
MIOSH - Member of the Institution of Occupational Safety and Health
NEBOSH - National Examination Board on Occupational Safety and Health
NVQ - National Vocational Qualification
PPE - Personal Protective Equipment
ROSPA - Royal Society for the Prevention of Accidents
SCOTVEC - Scottish Vocational Education Council

F. KEY HEALTH AND SAFETY LAWS

- Health and Safety at Work Act 1974
- Management of Health and Safety at Work Regulations
- Workplace (Health Safety and Welfare) Regulations
- Provision and Use of Work Equipment Regulations
- Personal Protective Equipment at Work Regulations
- Health and Safety (Display Screen Equipment) Regulations
- Manual Handling Operations Regulations
- Control of Substances Hazardous to Health Regulations
- Noise at Work Regulations
- Construction Regulations
- Construction (Head Protection) Regulations
- Electricity at Work Regulations
- Factories Act
- Fire Precautions Act
- Ionising Radiations Regulations
- Offices Shops and Railway Premises Act
- Offshore Installations Regulations
- Safety Signs Regulations